AMAZING ROBOTS!

PETER HOLLAND

ARGUS BOOKS

Argus Books
Argus House
Boundary Way
Hemel Hempstead
Hertfordshire HP2 7ST
England

First published by Argus Books 1990

© Peter Holland 1990

All rights reserved. No part of this publication may be reproduced in any form by print, photography, microfilm or any other means without written permission from the publisher.

ISBN 1 85486 041 0

Phototypesetting by Project 3 Filmsetters, Whitstable.
Printed and bound in Great Britain by Cambus Litho, East Kilbride, Scotland.

CONTENTS

INTRODUCTION	4
SWINGER	9
HI-HAWK	13
CABLE CREEPER	17
QUADRA	21
GLOAK	25
BIG CLAW	29

INTRODUCTION

IN this book you will find card sheets printed with the parts of six robot machines for you to cut out and glue together. Each is powered by a rubber band, and all use similar methods of propulsion.

You will need a few household items to make the simple mechanical parts. These are almost the same for each model, so once you have the easiest machine working, you can progress right through to the final fearsome-looking 'Big Claw'.

You only need simple tools. They are scissors, a modelling knife and (for bending and cutting paper clips) snipe-nosed pliers, Fig.A. Also useful are a pencil and a ruler.

To help you to complete the models, you can find the following items around the house: paper clips, pins, matchsticks, cotton bud stems (for bearings and washers) a balloon, rubber bands and UHU or PVA wood glue.

HOW DO THEY WORK?

If you twist a rubber band, by fixing one end and hooking the other into a crank or shaft (made from a paper clip), you are storing energy. When you let go, the crank spins round the other way. The rubber band and crank are fitted to a cardboard tube. This is the 'power tube'. The crank goes into a slot in another piece of folded card, called a 'crank box', which tilts from side to side, moving the legs, Fig.B.

The legs run on wheels, which have brakes to stop them skidding backwards. This makes the robot move forward, see Fig.C.

Although all the robots use this basic method of movement, each is excitingly different for you to enjoy making. The following chapter gives tips showing you how to use the card parts and put them together. Each model uses these methods, so refer back to them as you build, and includes a number of sketches to show you how to cut, fold and glue its own parts.

Build your six robots in the order in which they appear, and amaze your friends.

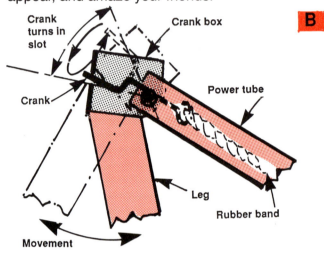

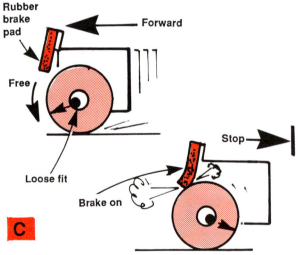

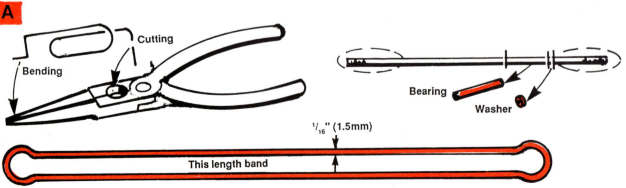

BUILDING TIPS

At the back of this book, you will find a set of card part sheets. You must cut all the parts out in sequence. The instructions in each chapter tell you when to cut them, and each part is identified by a letter beside the part.

Cut the card with scissors, so that the outline is on the edge of the cut. Do not cut inside the lines or on any dotted lines (Fig.1.). To prevent mixing up the parts, cut only one at a time and mark its letter on the back with a pencil. This way, if it shows later, it can be rubbed out (Fig.2).

Folding

Score the card *lightly* with a knife. Do not press down too hard, as the card only needs a slight break in the surface to make it bend. Use a ruler to guide the knife — all the lines are straight.

There are two types of line:
— 'dash - dash' (- - - -), for bending down, should be scored on the *printed* side of the card.
— 'dash - dot' (- · - ·), for bending up, should be scored on the *unprinted* side of the card.

When scoring a 'dash - dot' line, prick through the card at each end on the printed side and turn the card over to score the line from pinhole to pinhole.

The best way of bending long narrow tabs at the edge of the parts is to place a ruler on the tab and bend the rest of the part up (Fig.5).

Gluing

Use UHU or PVA (white wood glue) when joining one completed part to another. Look for a small identifying letter which will tell you where it is to be glued. If this letter is in a circle, then glue that part to the *printed* side; if it is in a square, then it should be glued to the *unprinted* side. For example, in Fig.6:

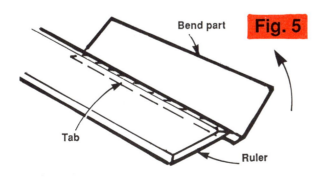

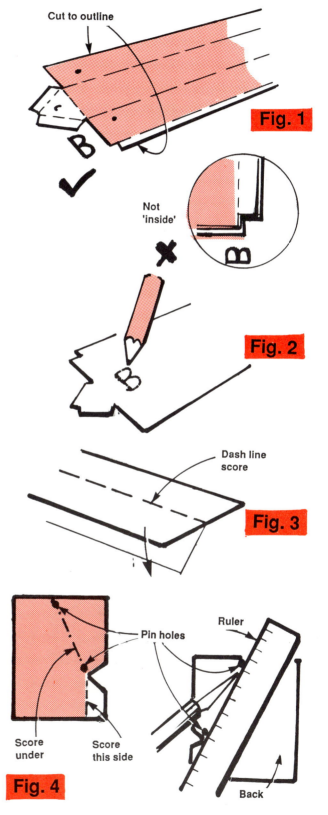

5

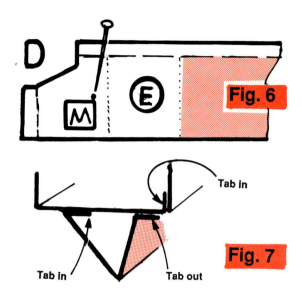

Fig. 6

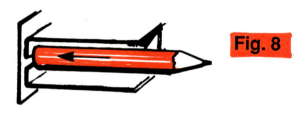

Fig. 7

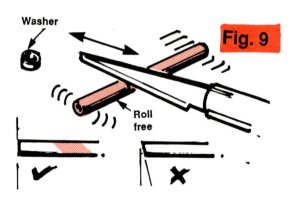

Fig. 8

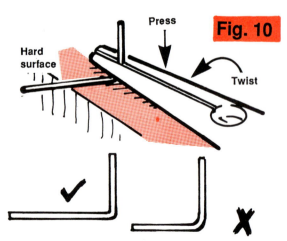

Fig. 9

Fig. 10

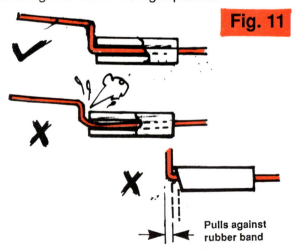

Fig. 11

— the small E means glue part E to the printed side here
— the small M means glue part M to the unprinted side here.

Gluing to the unprinted side is easy when the join is near an edge but, if it is near the centre of the part, prick through with a pin and mark in pencil on the back.

The construction sketches in the back show what a slice through the card should look like, and tell you which tabs go over and which turn under (see Fig.7). Press down any difficult-to-reach joints with a pencil before the glue dries (Fig.8).

Bearing tubes
A cotton bud stem makes a good tube. Take the cotton off the ends and you will see that there is a rough surface. Use this part to help the glue to grip when you make a crank bearing for the power tubes. Measure the length from the sketch in the instructions and cut by rolling the tube under the knife, as if you were sawing, but allowing the tube to roll. This parts it evenly. The end *must* be straight and not slanted (Fig.9).

The cranks
Straighten out a large paper clip and shape it with the pliers to a very sharp right angle. Press it on a hard surface as in Fig.10 because, if the bend is rounded, the crank will be stiff to turn in the bearing tube. It will also run unevenly if the tube is not cut straight (Fig.11).

Lay the shaped wire on the sketch so that it matches exactly. Adjust it with the pliers to suit. Do not bend the hooked end for the rubber band until you have pushed it into the bearing tube.

Wheels
The wheels *must* run true. It is a good idea (though not essential) to make a little 'jig' to help you to get the bearing and rim in the right places.

Look at Fig.12. Take a cork or piece of wood for a base. Using the cutting part of the pliers, cut the heads off 3 or 4 pins. Cut the cotton bud bearing tube to length — most of them are $5/16''$ (8mm) long. Slide the pins into it until no more will go in; this may take three or four. Push the points into the cork with the tube upright. You will be able to pull the tube off when the wheel is finished.

Scrape or slice a matchstick down to $1/16''$ (1.5mm) thick, and cut off two pieces $1/2''$ (12mm) long. Glue to the base each side and close to the tube. Cut off two more pieces each $1/4''$ (6mm) long and glue these across the first pair. When dry, this forms the jig.

To make the wheel, cut the card disc out with scissors so that the edge line is still there, and there are no bumps or flat bits. The hole for the bearing is square — it's easier to cut square holes with a knife. The tube will be a tight fit to press the hole to shape. UHU-type glue will do the rest. The disc should rest on the top matchsticks and be half-way up the tube. The wheel rim is a card strip. Pre-curve it by bending it over a pencil, then glue it to the disc edge so that it rests on the lower matchsticks. When dry, the wheel should spin freely on a single pin.

The wheel needs a tyre, so cut a slice of rubber from the neck of a balloon — use the narrow part, not the thick ring at the end. The slice should be the same width as the rim. Fix it to the rim with tiny spots of glue. It must be smooth. Wipe away any glue that gets on the outside as this will stop it gripping.

Winding the rubber band
You can wind most of the models by holding the power tubes near the crank box and turning the crank, as in Fig.13. The rest of the model will jerk from side to side, so it must hang downwards.

Alternatively, wind the rubber from the opposite (fixed) end. Make a wire hook from a paper clip, as in Fig.14. This replaces the matchstick normally used to hold the rubber band. It has a loop into which a second piece of wire will hook. This is the winder. Use it to pull the first clear of the end of the tube, then spin it between finger and thumb to wind the rubber. Hold the wire with the other hand between 'spins', or let the wire seat back on the tube end. Count the turns to avoid overwinding, and so breaking the band. It should stand about 150 turns.

Band replacement
With care, the band should last for a long time. A smear of washing-up liquid before installing it will also help. Should it break, however, you will have to

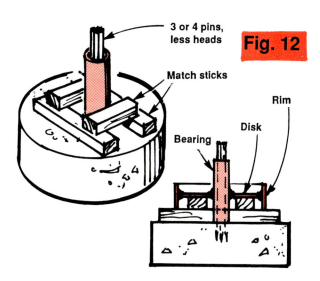

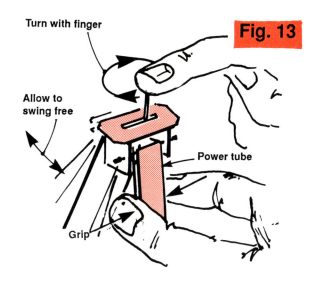

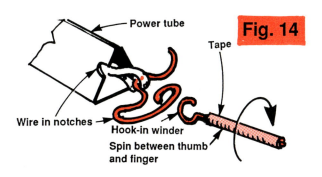

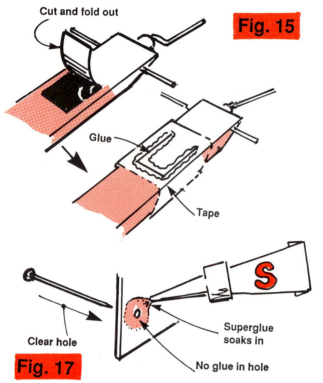

cut a small hatch in the power tube near the crank to get at the hook, before threading a new band in with a piece of wire. This will weaken the tube, which must be patched with masking tape after the hatch has been folded back and glued. Do this before winding fully again. Fig.15 shows you how.

If you cannot find thin rubber bands of the right length, loop shorter ones together, with the knot shown in Fig.16.

The holes where pins pivot may wear loose, but a tiny spot of superglue near, not on, the hole strengthens it. Clear the hole, only when set, with a pin (Fig.17).

How well the models work will depend on how carefully you make them, so start with Swinger, the easiest model, then enjoy the rest of your trip through *Amazing Robots!*

NOTE:
All the models described in this book have been built from the drawings and accompanying text, using only the recommended materials and assembly notes. The success of their operation will depend on the reader's adherence to the instructions and his/her individual ability.

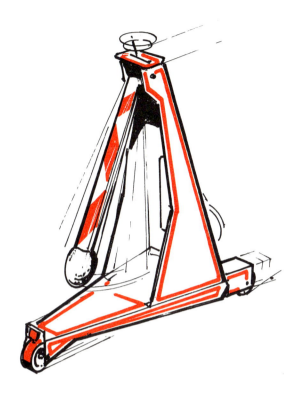

1 SWINGER

THIS machine is unusual in that although there is no connection between the power system and the wheels, it still moves forward!

The reason for this is that the power tube has a weight at the bottom end, like the pendulum of a clock. As there is more than enough power to keep it swinging, the excess power moves the machine backwards and forwards. However, to make sure it only moves forwards, a brake goes on each time the machine tries to travel backwards, but goes off when it tries to move forwards.

That weight does something else. It stops the rubber band unwinding too fast, so it is also a speed control! The weight of the swinging parts on the other robots in this book is used for the same purpose, otherwise they would shake and rattle and get nowhere in a short time, before needing a rewind. This slowing down mechanism gives them time to work properly and you time to enjoy them.

To build Swinger, take out the first of the card pages from the back of the book, then, using these instructions and the step-by-step sketches, make the parts one at a time. Check that each card part will fold into the right shape shown in the sketches by first folding it dry, then glue it together when it fits properly. You will make a mess of the card if you have to peel it apart to get it right. However, the first parts are from wire and tube, so get hold of your pliers.

Step 1 Bend the crank from a paper clip so that it matches the sketch.

Remember to leave enough wire at the other end for making a loop later. Cut a slice of cotton bud stem for the bearing, leaving the rough surface for better glue grip.

Slide the tube onto the wire with the smooth cut

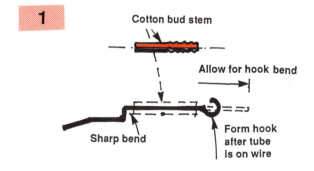

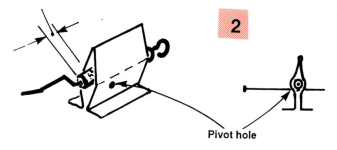
Pivot hole

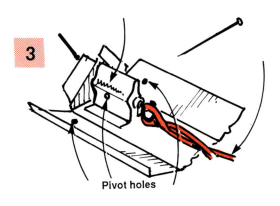
Pivot holes

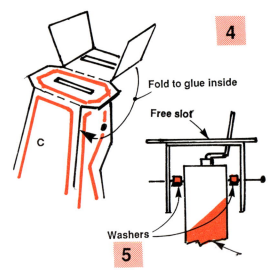
C — Fold to glue inside
Free slot
Washers

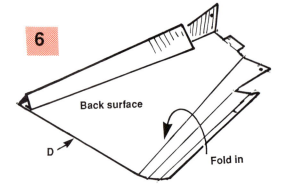
Back surface
D
Fold in

end next to the bend. Now, again with the pliers, bend the other end into a small loop to take a 5½" (140mm) long, 1/16" (1.5mm) thick rubber band, which you will fit later. Check that the wire spins freely in the tube.

Step 2 Cut part 'A' from the card, score the edges and the centre three lines on the printed face. Pierce the dots with a pin. Fold the card over the bearing with the crank end nearest the dot, but sticking out 1/8" (3mm). Glue the tube in and fold the edge tabs back so that they do not get glued together as well. Clamp with a clothes peg until set. Wax on the peg will prevent it sticking to the card.

Step 3 Cut out part 'B' and pierce, score and bend it to a 'U' shape, with the long side tab facing in. Remove the dry part 'A' with crank and apply glue to its tabs. Position it inside 'B' with the crank and tube end through the hole in the upturned triangular end. Before the glue sets, push a pin through the dots in the sides of 'B' and 'A' to align it. You will have to push the sides together dry, but allow them to open when you remove the pin.

When the glue is set, put the rubber band on the hook, lay it down inside 'B' and hold it temporarily with masking tape to the end. Glue in the long, and short, folded tabs to form 'B' into a triangular tube. This is best done on a flat surface to get the seam straight. The power tube is now complete.

Step 4 Cut out, pierce, score and fold part 'C'. This is the main frame and the crank box is formed out of the top end. Remember to cut the slots with the knife on a smooth surface, and to clear excess glue from them when the top is folded to make a double thickness.

Step 5 Cut two short pieces of cotton bud tube to form spacing washers. Pass a pin through the hole in one side of the crank box, and then through a washer. With the flat side of the power tube facing out, and the crank in the slot, the pin goes through the holes already in this part. Add a second washer and push the pin through the opposite side of the crank box hole. Secure the rubber band with half a matchstick at the end of 'B', and fit around it a piece of modelling clay about the size of a small marble.

Step 6 Cut out part 'D'. Score all long lines to bend down and the short one to bend up. Turn the part over and fold the edges inwards to make a triangular tube down each side. These stiffen it. Only glue the side tabs as the short one is to be used later for the brake pad. This part is the chassis and front axle box. Pierce the dots with a pin for the axle.

Step 7 With the bottom tabs of 'C' folded in, and the flat side of 'D' upwards, glue the two together. The sides of 'C' bend to match the shape of 'D'

Step 8 Cut out, score, pierce and fold the two axle boxes 'E' which go at the back. Glue these to 'C' so that they half overlap 'D'. They must both face straight back.

Step 9 Cut out the wheel discs 'F' so that they are true to the line. Cut the holes. Slice three pieces each $5/16$" (8mm) long from cotton bud tube for the bearings and glue these into each disc, with an equal amount sticking out each side. Make sure they are not sloping. When they are dry, cut the rim strips ('G') and glue them around the discs, While they dry, cut slices the same width from the neck of a small balloon to form tyres to grip the ground. Fix them with tiny dots of glue on the rim. Pass pins through the axle boxes with the wheels in place. The wheels should be loose on them.

Step 10 Cut a small piece of rubber from a $1/4$" (6mm) wide elastic band. Roll the front wheel forward and glue rubber to the upturned tab so that it touches the tyre where shown in the sketch. Let the wheel slip back, so that it spins free. Adjust by bending the tab until the wheel locks when the machine tries to go backwards, but is free to go forwards. You have now made Swinger!

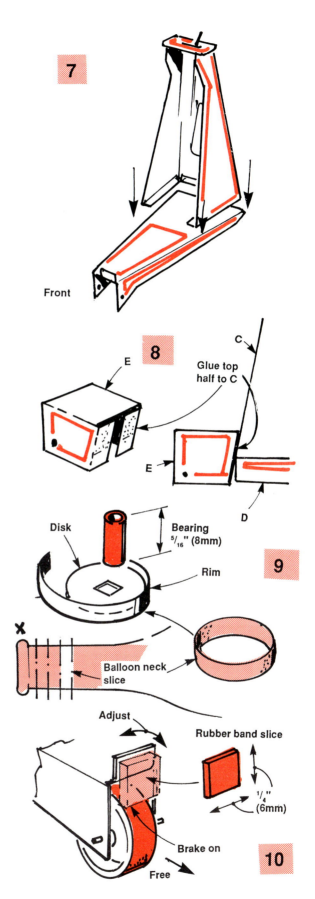

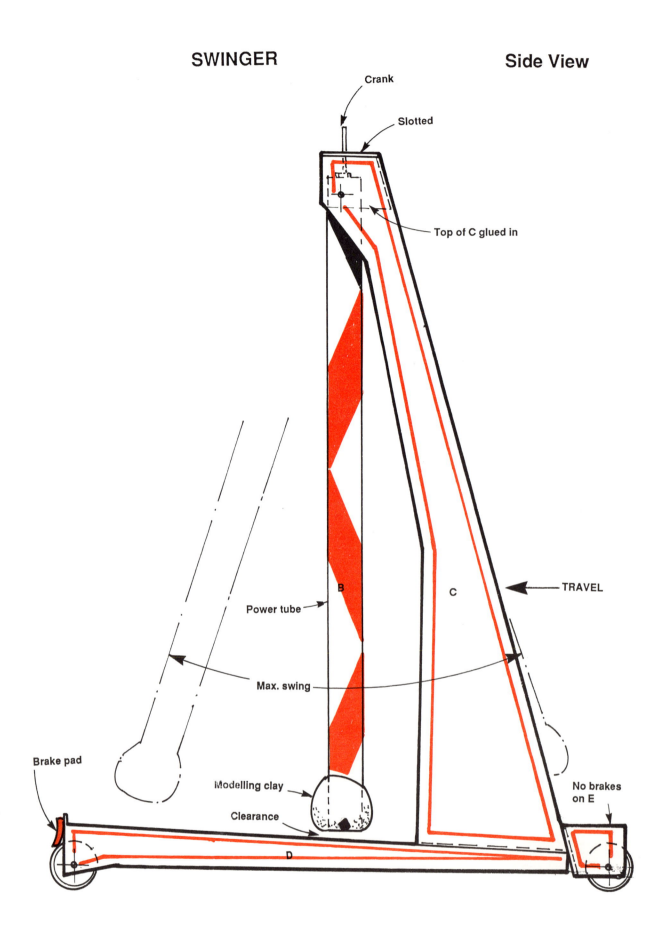

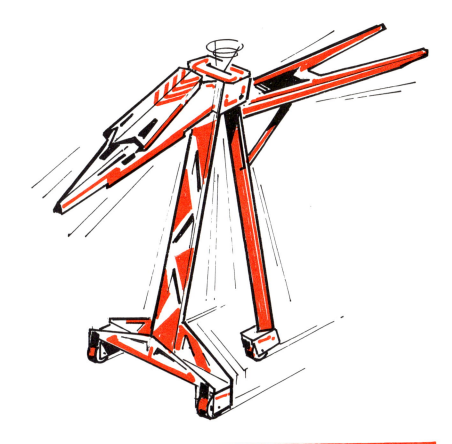

2 HI-HAWK

THIS robot has two legs, one behind the other. A power tube forms one leg and the other leg is fixed to a crank box. It walks with a pincer-like movement on three wheels. Each wheel has a brake which comes on if the leg tries to skid backwards. This keeps the machine moving forwards.

First remove the card sheet of Hi-Hawk parts from the back of the book. Take care to cut clear of the parts when doing this. Look at the step-by-step instructions and sketches as you build.

Step 1 Take a cotton bud stem, remove the cotton and, using the rough end for better gluing, cut a slice ¾" (18mm). Make sure the cut ends are smooth. Using snipe-nosed pliers, bend the crank shape from a paper clip. Check you have done this properly by laying it on the sketch. Slide it into the tube, then bend the other end to make a hook for the rubber band.

Step 2 From the card sheet, cut out part 'A' and score the dashed lines on top and the dash-dot lines from below. Glue the tube between the folded sides, so that the end with the crank projects ⅛" (3mm) at the pierced end. Clamp it with a clothes peg while the glue sets.

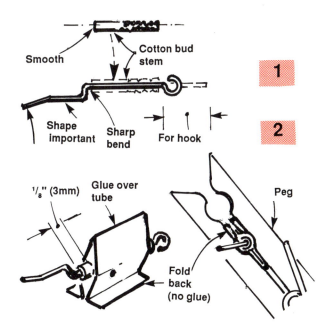

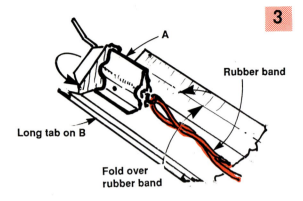

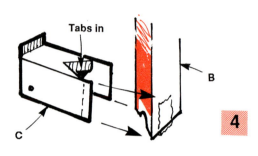

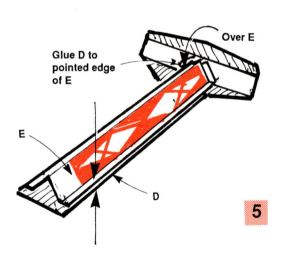

Step 3 Cut out, score and pierce the power tube part 'B'. Bend to form a 'U' section, then insert part 'A' and align it just as you did with Swinger, not forgetting the rubber band.

Step 4 Cut out card part 'C'. Score and fold it as shown, and glue it to the lower end of part 'B'. This is the axle box for the rear wheel.

Step 5 Cut out part 'D'. Score and bend down to make a 'U' shape at the wide end. Cut out part 'E', score it down the centre, turn it over and score the edges and pointed end. Fold the sides and end tabs up and fold the centre to make a 'V' shape. Glue it *under* part 'D'. This forms the front leg.

Step 6 Cut out the two axle boxes parts 'F', score to fold down as shown and glue under the ends of the front leg, so that the back of each box rests against the back of the leg.

Step 7 Cut out card crank box 'G'. Note that two chain lines on the right-hand end should be scored underneath to bend *up*. Cut the slots, and fold and glue so that the right hand end fits inside the left hand end, with the slots matched up. Clear excess glue from the slot. Glue the sloping end to the rear of part 'D', with the slotted face next to the end.

Step 8 Hold the power tube 'B' with its flat face outward and the crank in the slot in the crank box. Pass a pin through the holes in the crank box and those in the power tube, with small spacers each side of the tube 'B'.
   The bottom end of the rubber band should go over a half matchstick. Hold the power tube upright with the front leg hanging free. Wind on 20 turns.

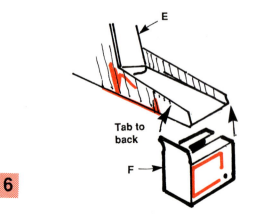

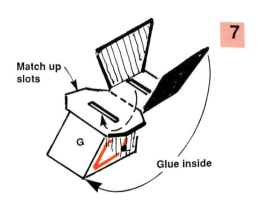

Let the matchstick seat in the notches in the lower end. The leg should swing to and fro as the crank turns.

Step 9 Cut, score and fold parts 'H' and 'I' to form the control cabin and roof. Glue 'H' to the front of 'E' and the tabs of 'I' onto 'H' where shown.

Step 10 Cut, score and fold part 'J'. Note that the strut at the right folds to form a 'V' section to make it stiff. Fold this under as a support and glue the whole to the rear of part 'B', as shown.

Step 11 All that remain are the wheels. Parts 'K' and 'L'. Make them the same way as for Swinger. When the glue is almost set, test spin the wheels on a paper clip and adjust until true. Allow to harden.

Step 12 Pass a pin through the points marked on the axle boxes and through the wheel bearing tubes. The wheels are a loose fit on the pins, so that they roll clear of, or up to, a scrap of rubber band which you now glue to the top tab of box 'C' and front of 'D'. These are the brake pads, like the one on Swinger. Check that they stop each wheel when the leg is dragged back, but are clear when it goes forwards. Bend the card to adjust.

Winding
Hold the rear (power tube) leg near the top, with the front leg vertical. Turn the crank to wind up the rubber band about 150 turns and place Hi-Hawk on a smooth level surface. If you have followed the instructions correctly, it should walk with a stride of about 2" (50mm). The diagram on the next page shows how to check it out. If the step is uneven, ie. slow to close, add tiny weights of modelling clay to the nose and fin tips.

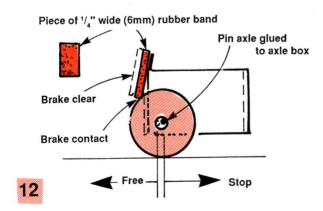

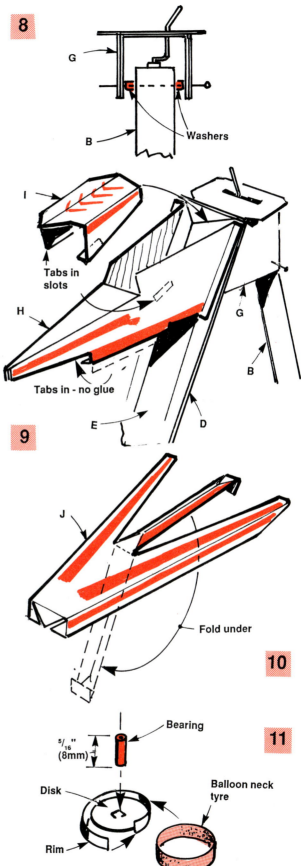

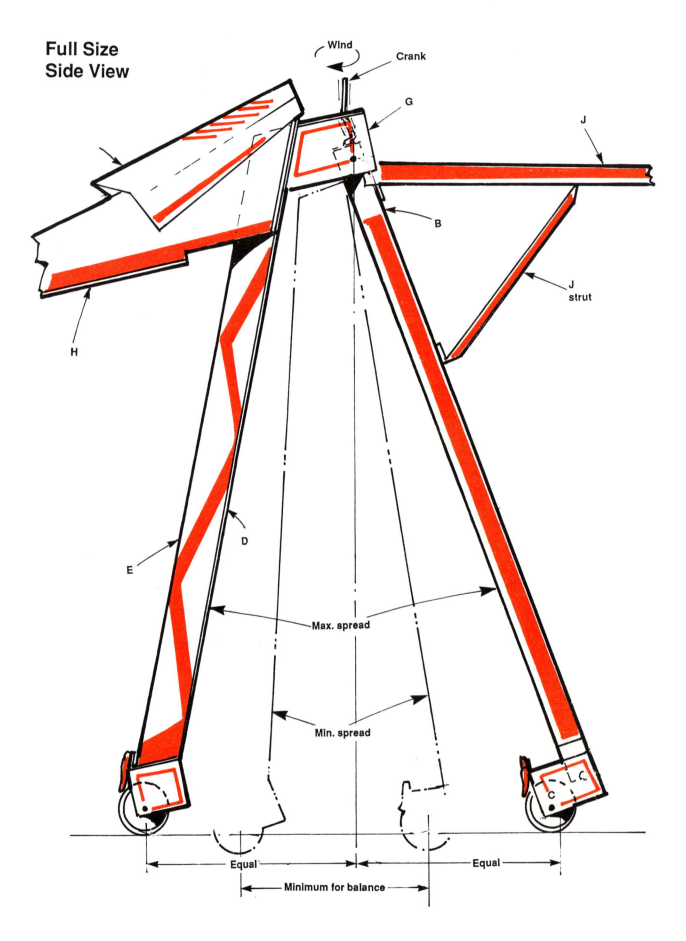

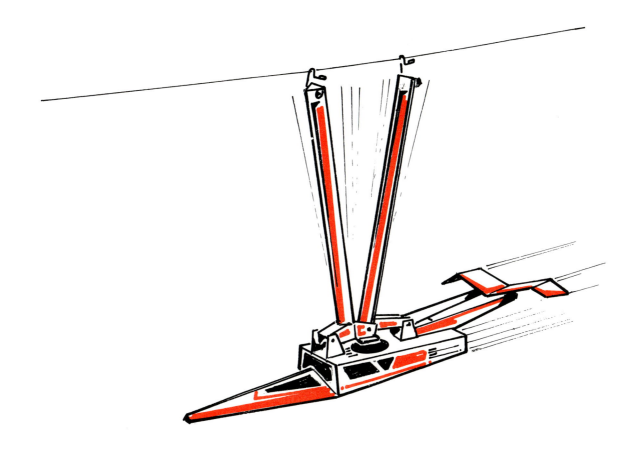

3 CABLE CREEPER

'CREEPER' is, perhaps, not the right word. This one glides along, while hanging below its supporting cable, which is a stretched piece of nylon monofiliment fishing line, or 'invisible' nylon mending thread. It uses the same 'pincer'-like leg movement as 'Hi-Hawk', but the legs are upside down. It slides on wire hooks instead of wheels and, of course, looks rather different. The hooks have rubber brake pads on them and they slide smoothly on the shiny nylon thread.

Step 1 Make the crank as in previous models. Do the same with the bearing tube.

Step 2 Cut out part 'B'. Note the extra holes at the other end, compared to earlier versions. Fit part 'A' in the same way and add the rubber band before closing up the tube.

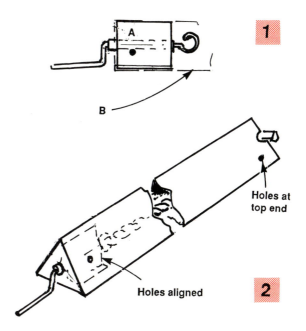

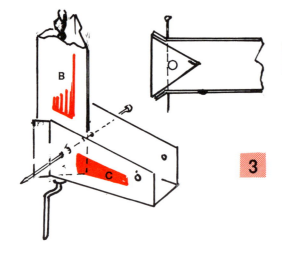

Step 3 Cut, pierce, score and fold part 'C'. Glue its tabs where shown on part 'B'. Push a pin through the holes and those in part 'B', so that the two are aligned on what will later become a pivot pin for the crank box — see the sketch.

Step 4 Part 'D' is like 'B'. Form it into a triangular tube in the same way.

Step 5 Cut, slot, pierce, score and fold the crank box, part 'E'. Note that one half is scored on the plain side, to fold inside the coloured half, with the slots matching up. Glue this double thickness and glue the finished box to part 'D', with the slotted part level with the end of 'D'.

Step 6 Make part 'F' in the same way as 'C'. Glue it to 'D' opposite to 'E', as shown in the sketch. Now combine the power tube with the crank box using a pin and tube washers, as with earlier models.

Step 7 The gondola is part 'G', and will hang from brackets 'C' and 'F'. Fold it up, but do not glue it, for it has to be shaped around part 'H' which you now make and glue to one side. Follow the numbered sequence of bending and gluing, otherwise it will not fit. Finish with the top of 'H' folded back on the top of the roof, with its arms upwards (check with the final drawing on the next page).

Step 8 Make part 'I', which is the rear support arm

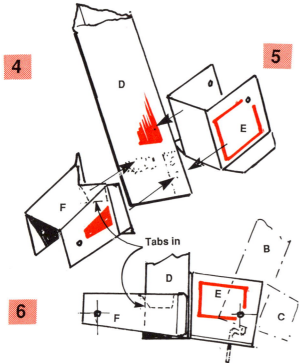

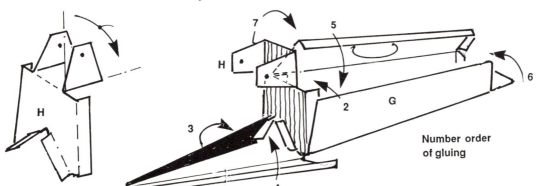

Number order of gluing

and spoiler outrigger. Glue it to the rear part of the roof. Cut, score and fold the spoiler 'J' and glue it on the rear of 'I'.

Step 9 Pass pins through 'C' and 'F', and through the slots in 'I' and holes in 'C'. The slots allow for movement when 'B' and 'D' spread or close.

Step 10 Using pliers, bend paper clips to the shape shown in the sketch. They have to pass through the top ends of 'B' and 'D' and hook over a nylon stretched tightly between a couple of chairs.

The shape is important, because the end carries a piece of ¼" (6mm) rubber band (the brake pads). This has to be close to the underside of the nylon. Glue a scrap of card on the other end of each wire where it passes through the ends of 'B' and 'D'. They have to move freely. Hang the model on the nylon and adjust the wires so that, when they lean back, the pads drop away from the nylon but, when they lean forward, they press against the nylon.

Wind up the elastic and test for action. If the adjustment is correct, the legs 'B' and 'D' should slide forward in turn without skidding backwards. Notice that the gondola does not bounce up and down or rock, due to the opposing movement of the levers 'C' and 'F'. Do not try to make the machine go up or down slopes. It has to hang from a level cable so that the wire brake hooks can work properly. So keep that nylon tight.

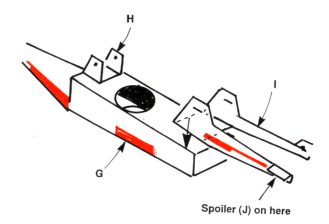

Spoiler (J) on here

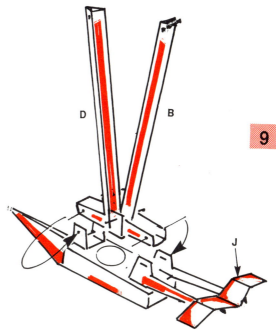

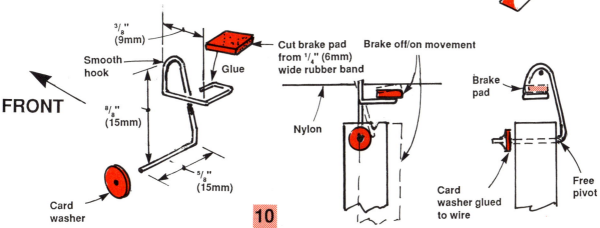

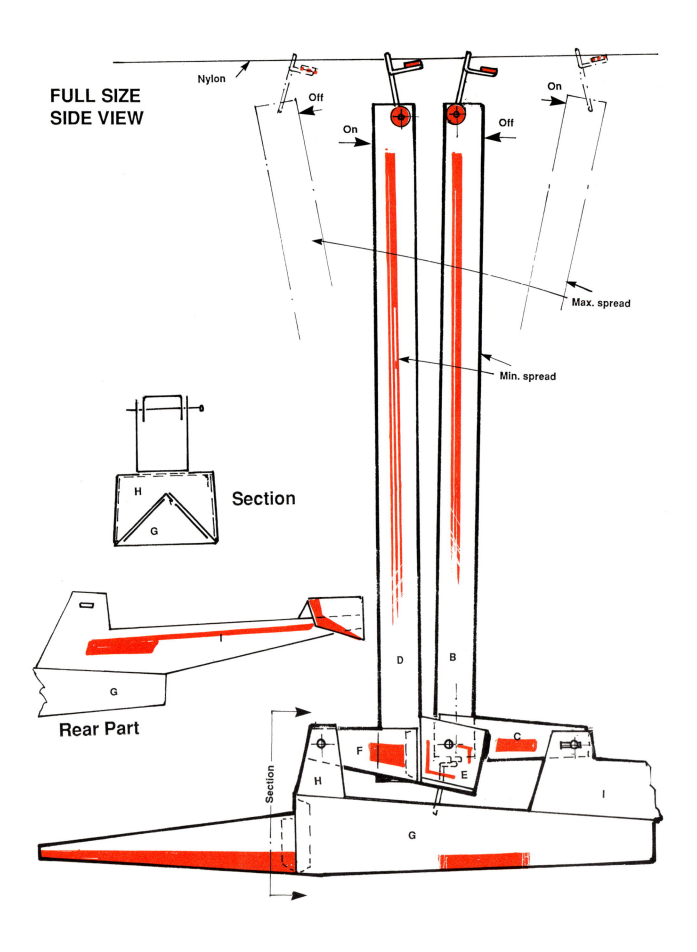

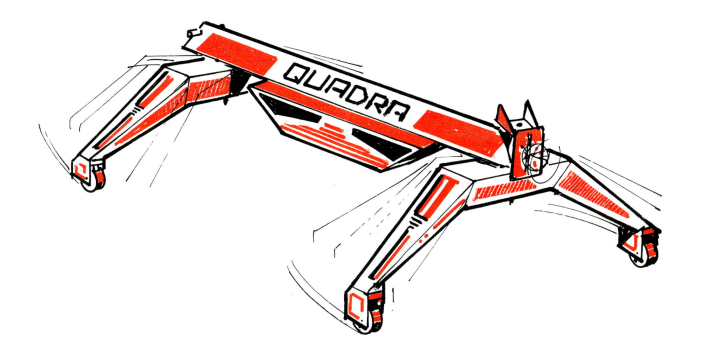

4 QUADRA

NOW for a four-legged machine! The Quadra swings its pairs of legs in unison, left, right, left right, front and rear, like an army on the march. They move in a semi-circular direction on vertical pivots, so that, as the left goes back, the right goes forward. Each foot or axle box has a wheel, and the front pair, or both, have automatic brakes just as Swinger and Hi-Hawk. There is no rise and fall in this system, so little power is wasted. First, the left foot moves, pivoting on the right, then the right foot pivoting on the left. Such action is known as the 'walking beam' and gives the rest of the machine a slight sideways swing with each step.

Step 1 Make the wire crank and bearing tube as in other machines.

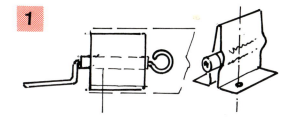

Step 2 Make part 'A', but note that there is a hole in one of the tabs this time. This is because the pivot pin has to be vertical instead of horizontal.

Cut out and make part 'B' as before, and install 'A' with a pin through for alignment. Note that the pin holes are in different places and at both ends. These are for vertical pivots.

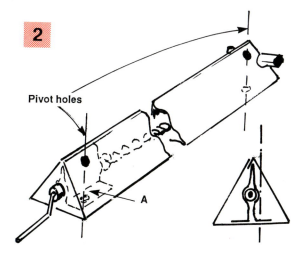

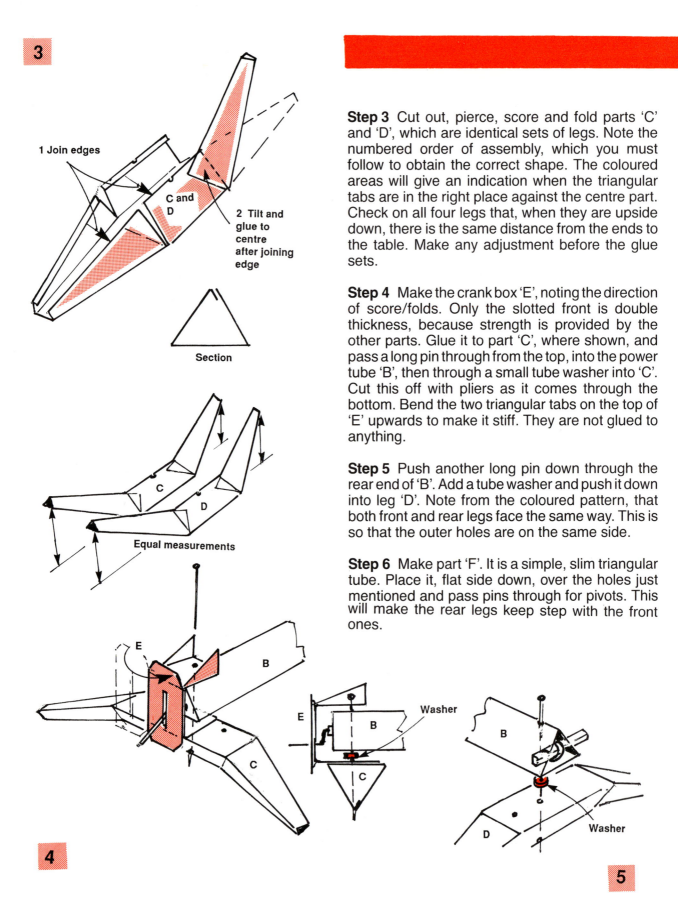

Step 3 Cut out, pierce, score and fold parts 'C' and 'D', which are identical sets of legs. Note the numbered order of assembly, which you must follow to obtain the correct shape. The coloured areas will give an indication when the triangular tabs are in the right place against the centre part. Check on all four legs that, when they are upside down, there is the same distance from the ends to the table. Make any adjustment before the glue sets.

Step 4 Make the crank box 'E', noting the direction of score/folds. Only the slotted front is double thickness, because strength is provided by the other parts. Glue it to part 'C', where shown, and pass a long pin through from the top, into the power tube 'B', then through a small tube washer into 'C'. Cut this off with pliers as it comes through the bottom. Bend the two triangular tabs on the top of 'E' upwards to make it stiff. They are not glued to anything.

Step 5 Push another long pin down through the rear end of 'B'. Add a tube washer and push it down into leg 'D'. Note from the coloured pattern, that both front and rear legs face the same way. This is so that the outer holes are on the same side.

Step 6 Make part 'F'. It is a simple, slim triangular tube. Place it, flat side down, over the holes just mentioned and pass pins through for pivots. This will make the rear legs keep step with the front ones.

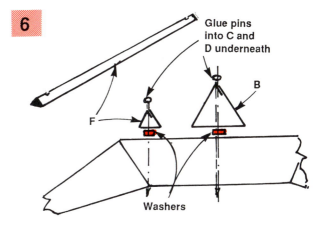

Step 7 Note that there are left and right axle boxes. To avoid muddles, make the 'G' types first and fix them to the *right* legs. Glue the long *side* tab to the top surface of the leg, and the short tab to the pointed lower edge. Do not confuse the longest part, which is the *top* of the box, including the brake pad support, in the centre. The open side of each box must face the front. Now make and fix the left axle boxes. parts 'H'.

Step 8 Make up four wheels from parts 'I' and 'J'. Don't forget to fit the balloon neck tyres.

Mount each on a pin in the axle boxes. Fit brake pads to the front two wheels only, and adjust them carefully until they give the correct freedom for forward movement and resistance for backward thrust. Wind the elastic and test the model on a smooth level surface. If the brakes grip well, there should be no need to fit the rear ones. Those rear legs are for stability, but will give four-wheel drive if brakes are fitted. You may find that it goes faster in two-wheel drive.

Step 9 Finally, make the gondola, 'K'. Glue the corner tabs inside the sides to keep the shape, then fold the sides in at the top edge and glue them along the line of colour to the flat bottom of the power tube 'B'.

Always run this machine on a flat surface. Lumps and humps cause the wheels to slip when the brakes should grip. With the exception of the Cable Creeper, *all* these machines work best on smooth level surfaces.

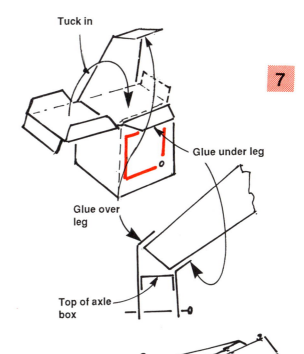

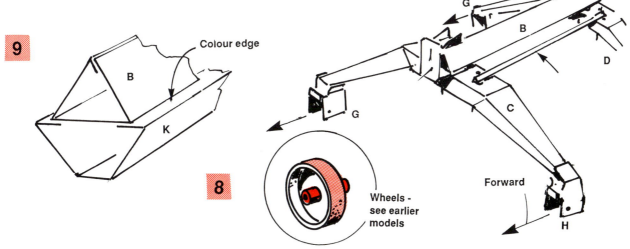

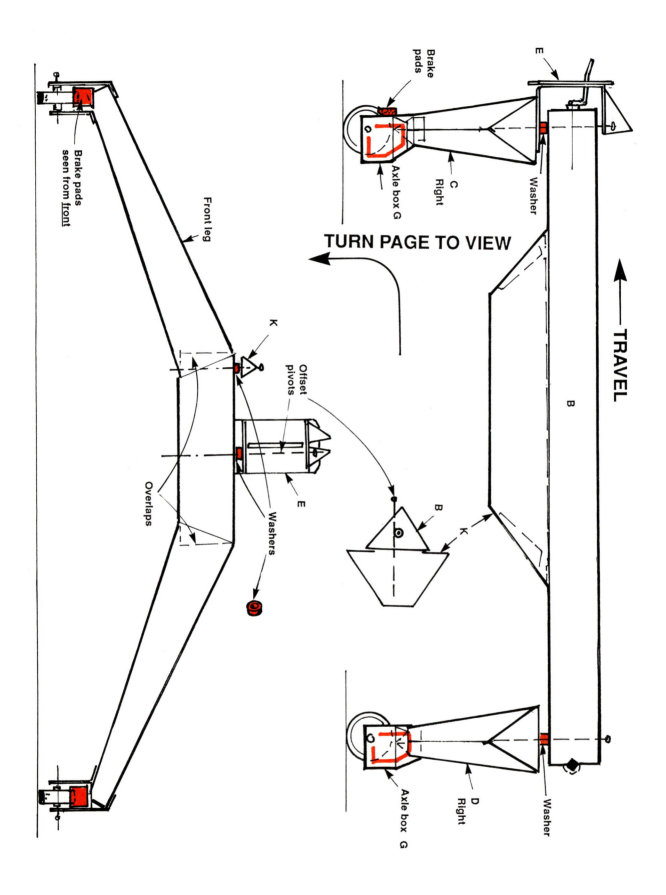

5 GLOAK

GLOAK (with the cloak) is a humanoid type walking on two legs and swinging his outstretched arms. He is another walking beam machine, and is kept upright by the cloak. This runs on a third wheel and is weighted so that it grips the ground to prevent sideways movement.

The power tube forms one arm. It is outstretched so that it pivots vertically in a crank box fixed to the cloak. The card parts are on two sheets at the back of the book. Detach them and start as before.

Step 1 Make the wire crank and bearing tube, as before. The end of the crank is a little shorter, because the machine cannot be wound by it. Cut and fix part 'A' to the tube.

Step 2 Make part 'B', which is the power tube arm, with 'A' inside. The flat side will be the front. Cut out the left hand 'C' and glue it to the end where shown.

Step 3 'D', a simple triangular tube, is the right arm. Cut out and glue the other hand 'E' to the end.

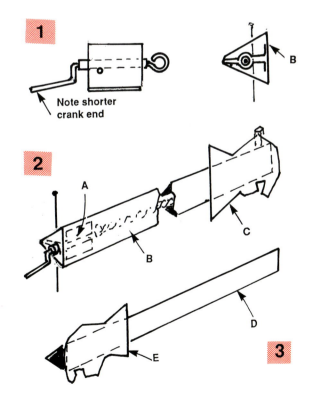

25

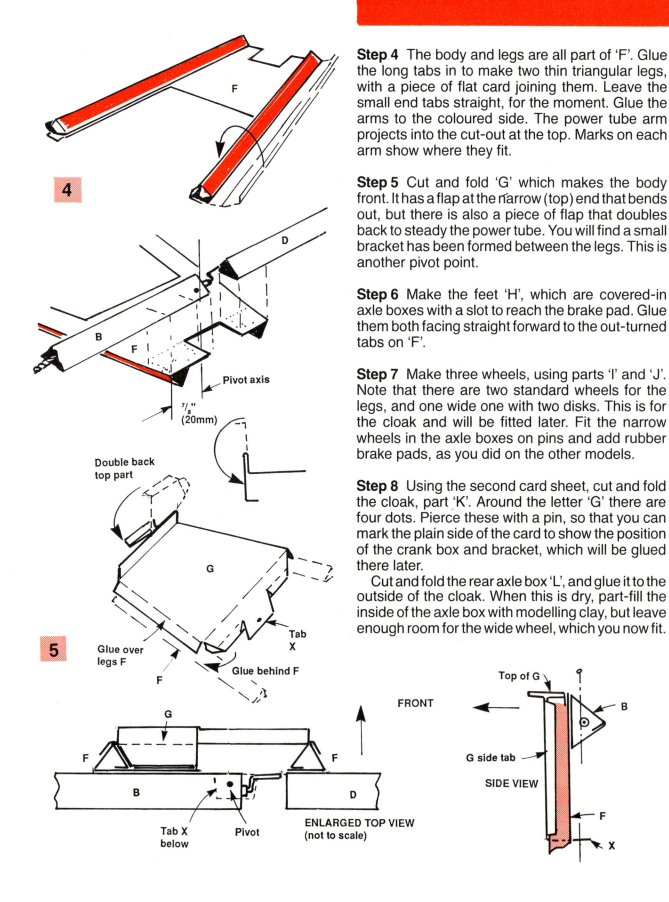

Step 4 The body and legs are all part of 'F'. Glue the long tabs in to make two thin triangular legs, with a piece of flat card joining them. Leave the small end tabs straight, for the moment. Glue the arms to the coloured side. The power tube arm projects into the cut-out at the top. Marks on each arm show where they fit.

Step 5 Cut and fold 'G' which makes the body front. It has a flap at the narrow (top) end that bends out, but there is also a piece of flap that doubles back to steady the power tube. You will find a small bracket has been formed between the legs. This is another pivot point.

Step 6 Make the feet 'H', which are covered-in axle boxes with a slot to reach the brake pad. Glue them both facing straight forward to the out-turned tabs on 'F'.

Step 7 Make three wheels, using parts 'I' and 'J'. Note that there are two standard wheels for the legs, and one wide one with two disks. This is for the cloak and will be fitted later. Fit the narrow wheels in the axle boxes on pins and add rubber brake pads, as you did on the other models.

Step 8 Using the second card sheet, cut and fold the cloak, part 'K'. Around the letter 'G' there are four dots. Pierce these with a pin, so that you can mark the plain side of the card to show the position of the crank box and bracket, which will be glued there later.

Cut and fold the rear axle box 'L', and glue it to the outside of the cloak. When this is dry, part-fill the inside of the axle box with modelling clay, but leave enough room for the wide wheel, which you now fit.

Step 9 Make the crank box 'M' and glue it inside the top of the cloak, with the narrow side up and level with the top edge of the cloak. Make the bracket 'N'. Fold the edges to make it rigid and glue it inside, level with the lower pin holes. The end tab doubles back to steady a pivot pin.

Step 10 Put the cloak onto the main part of the robot. To do this, slide the crank into the crank box slot, and push a pin through the crank box holes and through the holes in the power tube. The cloak should now swing from side to side as the crank turns.

There will be a lot of strain on this pin, so push a second pin up through the bracket 'N' into the one under 'F'. Fix its point to the back of 'F' with UHU or masking tape, to stop it falling out. Before adding Gloak's head, check that the crank turns freely. Wind the rubber from the hand end, and test on a smooth surface. The cloak should move straight forward, while the whole body-arms-and-legs unit twists from side to side.

If the brakes work as they should, Gloak will walk forward. If the cloak skids from side to side, yet the wheels free, add modelling clay at the back. You may need to adjust the crank shape to increase or reduce the amount of twist.

If Gloak behaves properly, make and fit his head in the next step.

Step 11 Cut out the cowl and face, parts 'O' and 'P'. Fold and curve 'O' and glue the sides to the top. Then bend back the tabs of 'P' and glue it inside the front edge. Glue the bottom edge of the cowl to the outside top edge of the cloak 'K'. The face will partly cover the crank and box, but may be bent back if you need to reach them.

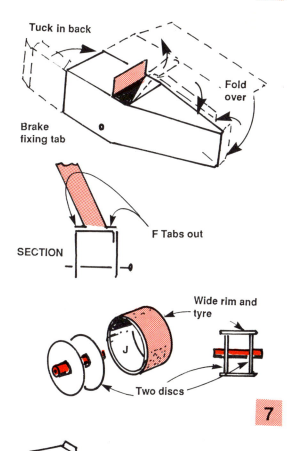

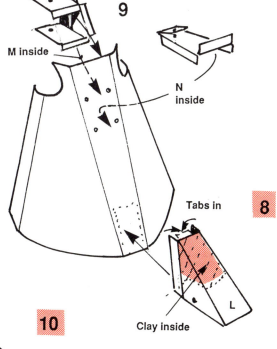

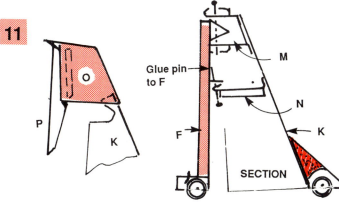

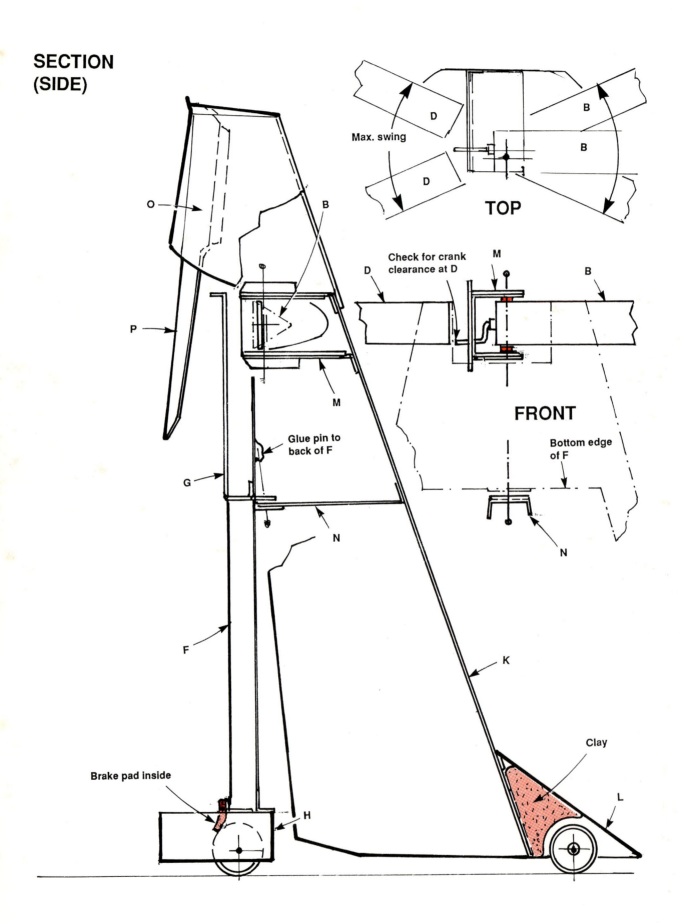

6 BIG CLAW

IT looks rather like an animal, but this robot-type demolition machine punches its way forward on a hidden walking beam. A rear support wheel is disguised by dummy hind legs, which move in the opposite direction to the claws, as it turns its head and swings its tail fins.

Step 1 Take out the last two card pages and make yet another crank, bearing and support piece 'A'. Cut out and assemble the power tube 'B' around this first part, remembering the rubber band and noting that the pivot pin will be vertical. There are also holes at the rear end for the hind legs.

Step 2 Make part 'C'. This is the walking beam, with the crank box attached. There are pivot holes at the centre and ends, and the flat side will be on top.

Make the two axle boxes 'D' and glue them under the ends of 'C' with the open ends facing forward, like the crank box. Glue the angled tabs under the beam and the long tab over the top surface. Pierce the end holes of 'C' again, if they become covered in glue or bits of tab.

Step 3 Make the rear axle box 'E' and glue it under the rear end of the power tube 'B', which has its flat

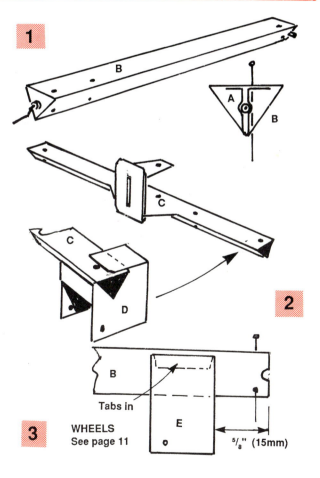

side on top. Make the three wheels from parts 'F' and 'G'. The one with the wider rim goes in the rear box. Fit each wheel on a pin axle and add rubber brake pads to the front pair. These go on the short front pieces of the boxes.

Step 4 Assemble the power tube to the crank box, with the crank in the slot and a pin through the box and end of the tube. Remember that the tube is flat side upwards. Make part 'H', which is a triangular tube that will be used to keep the claws straight. It forms a parallelogram with the beam 'C'. Pivot this on another pin through the power tube, but clear of the rubber band inside. 'H' should be flat side up, below 'B' with the inner hole on the right.

Step 5 Make the two claws. 'I' is the left and 'J' the right. Pass short pins through the front hole of each into the ends of the beam 'C', and through their rear holes into the ends of 'H'.

Cut out and form 'K' into a 'V' shape. This is a pushrod that will drive the hind legs. Pin it to the inner hole of 'H'.

Step 6 From the second card sheet, cut and assemble piece 'L' to form the hind legs. It will have a pivot bracket at the back. Leave this part unfolded at the moment.

Cut out and fold part 'M', which is a strengthening piece, 'V'-shaped in section, with the ends tapered down. Glue this inside 'L' (on the plain side). It will make the legs bend down.

Fold up the bracket part of 'L', fit it over the end of 'B' and pivot it on a pin. Now connect the leg with a pin in the right-hand side to the pushrod 'K' which comes diagonally under the power tube 'B'.

Step 7 Make the tail piece 'N' which has to be bent

to a 'W'-section at the end, and an inverted 'U'-section at the front.

Cut out 'O' which is a support bracket, and pivot it on a second pin in bracket 'L'. The pin must clear the end of the power tube. Glue the tail tabs to the vertical face of 'O'. Check that the tail swings when the hind legs jerk from side to side, but is not rigidly connected; some looseness is needed here.

Step 8 Make the head from parts 'P', which is the main section, and 'Q', which fits into slits in the top. Glue the tabs at the wide end to the top of the beam 'C' each side of the crank box. The head is open underneath, so that the crank can be reached for winding.

Step 9 Check that all moving joints are free, but not so sloppy that parts become entangled with each other. Hold the power tube vertically with the head up. Wind on about 50 turns. As you do so, keep clear of the claws and legs which will be moving to and fro. Now place the machine on a smooth level surface. If you have set the brake system up correctly, Big Claw should crawl forward.

Step 10 Cut out and shape 'R' which is the bodywork. The wide end goes to the front and must clear the head as it moves. Make the two support brackets, 'S', and glue them front and rear to the top of the power tube, where shown. Glue the bodywork to the brackets so that the ends clear the head and tail and the edges clear the beam, legs and pushrod. Bend the body to suit. Reach inside it to hold the power tube when winding. If you want Big Claw to move more slowly, add modelling clay to each end of the beam 'C' near the axle boxes, but clear of other moving parts.

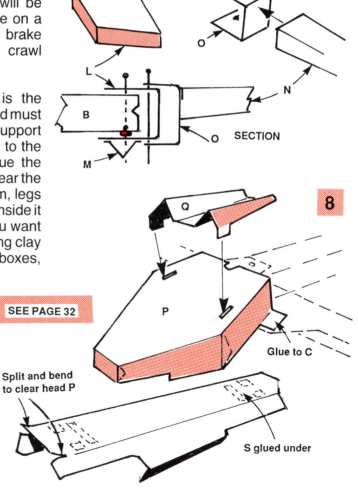

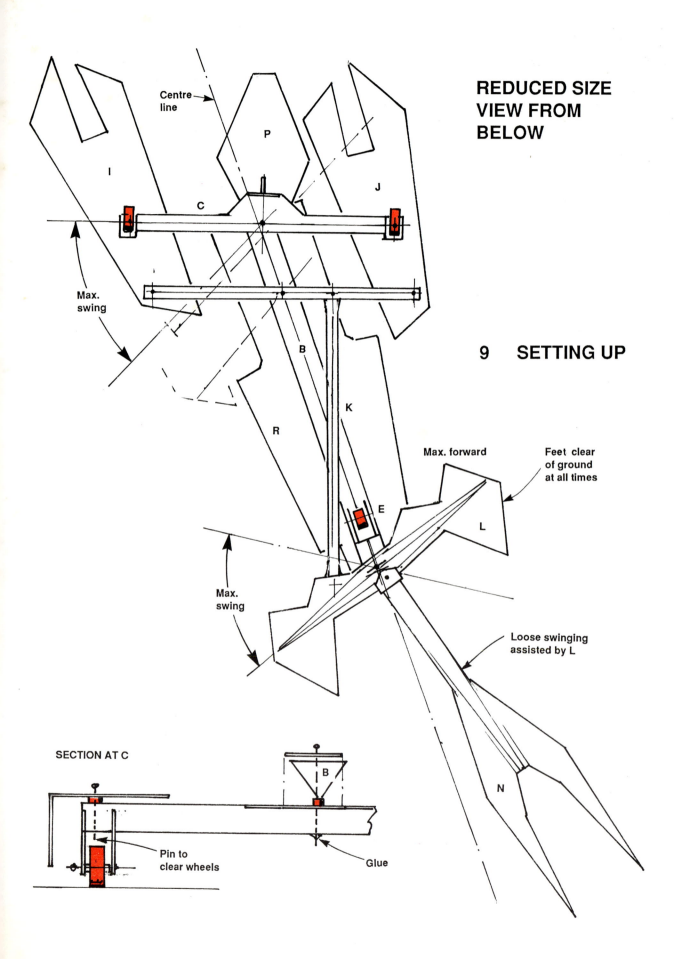

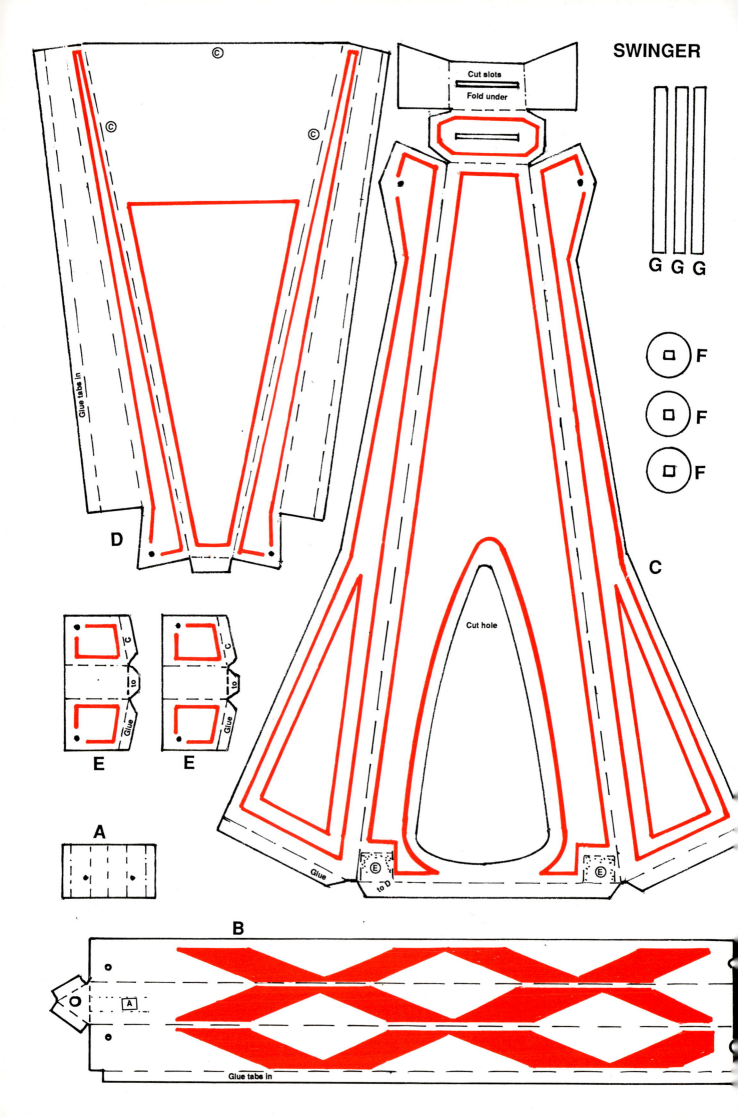

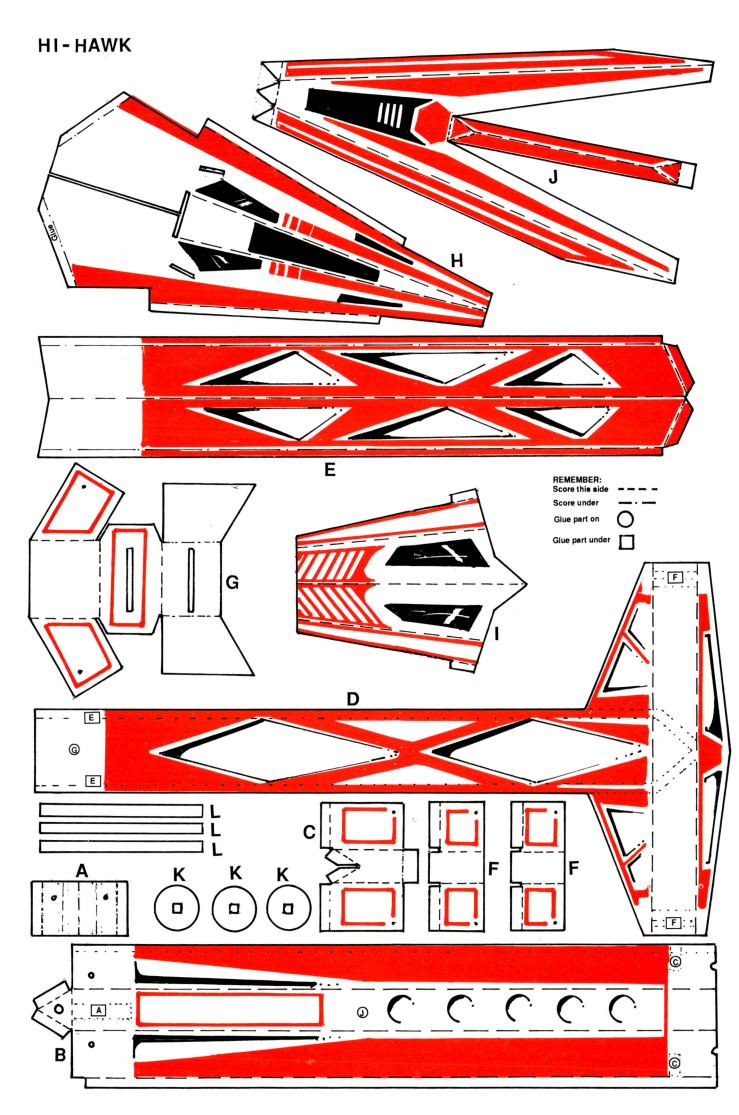

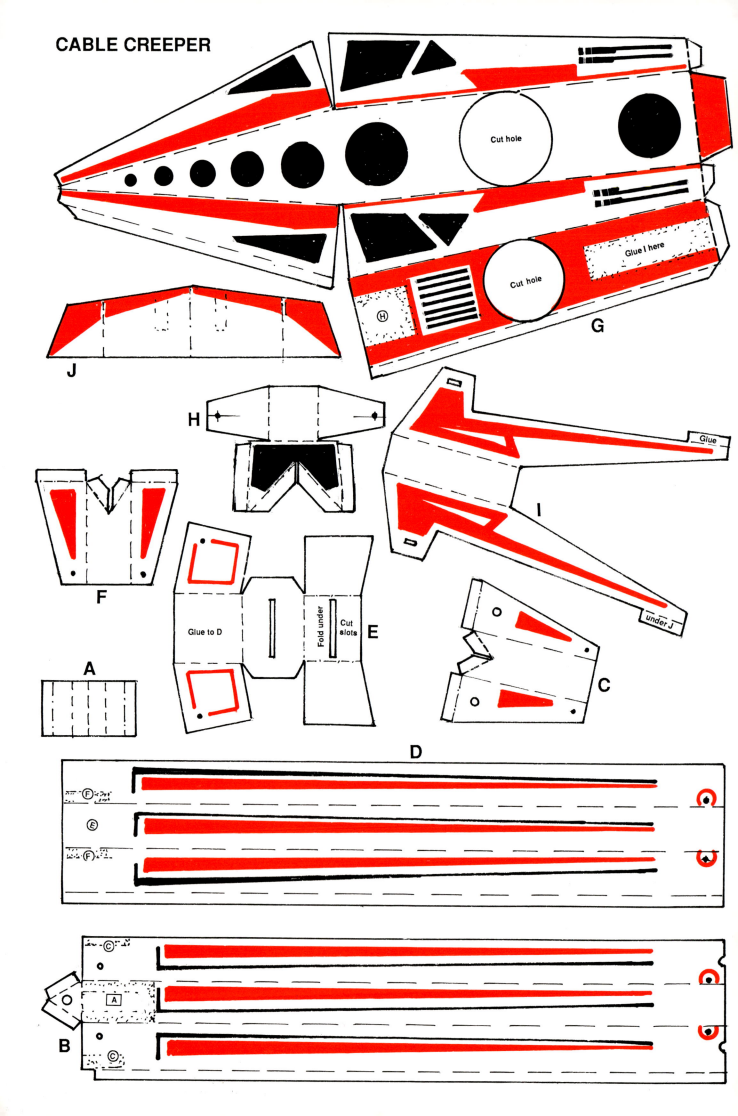

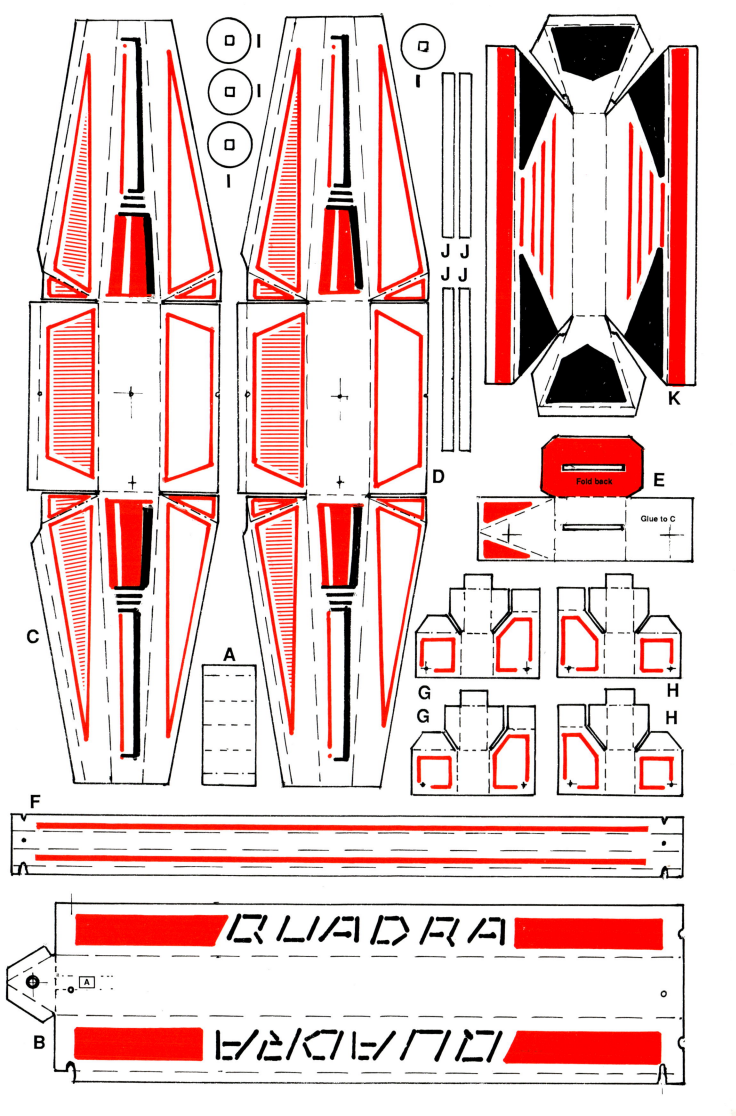

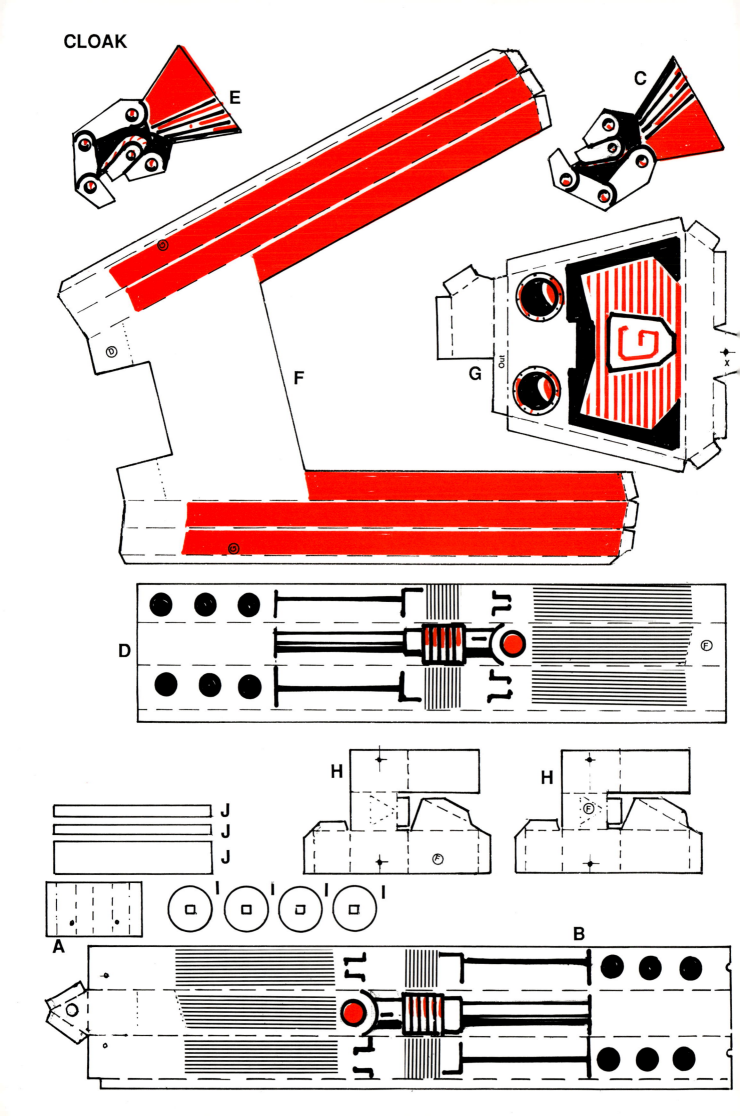

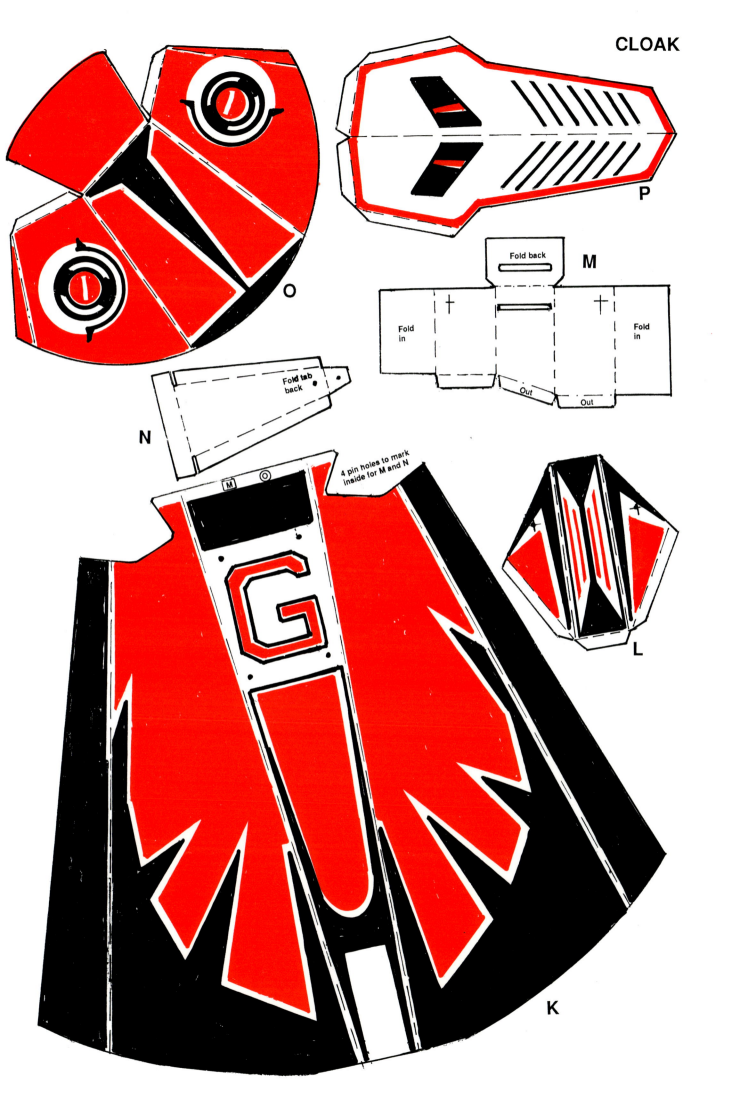

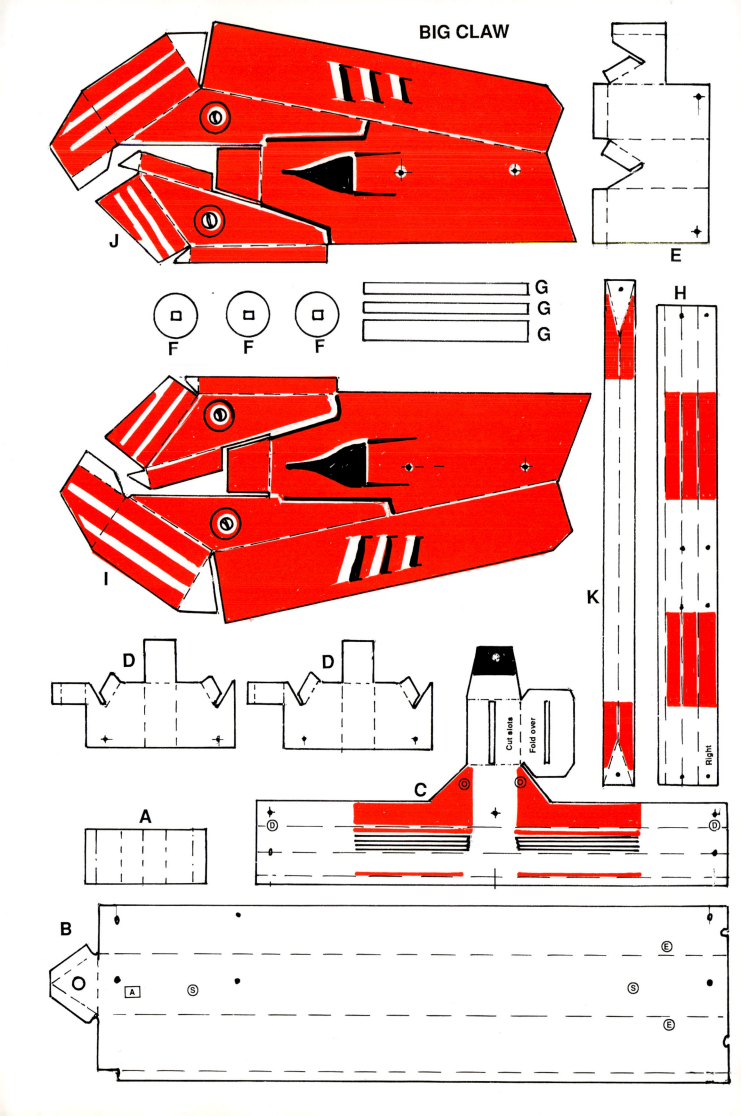

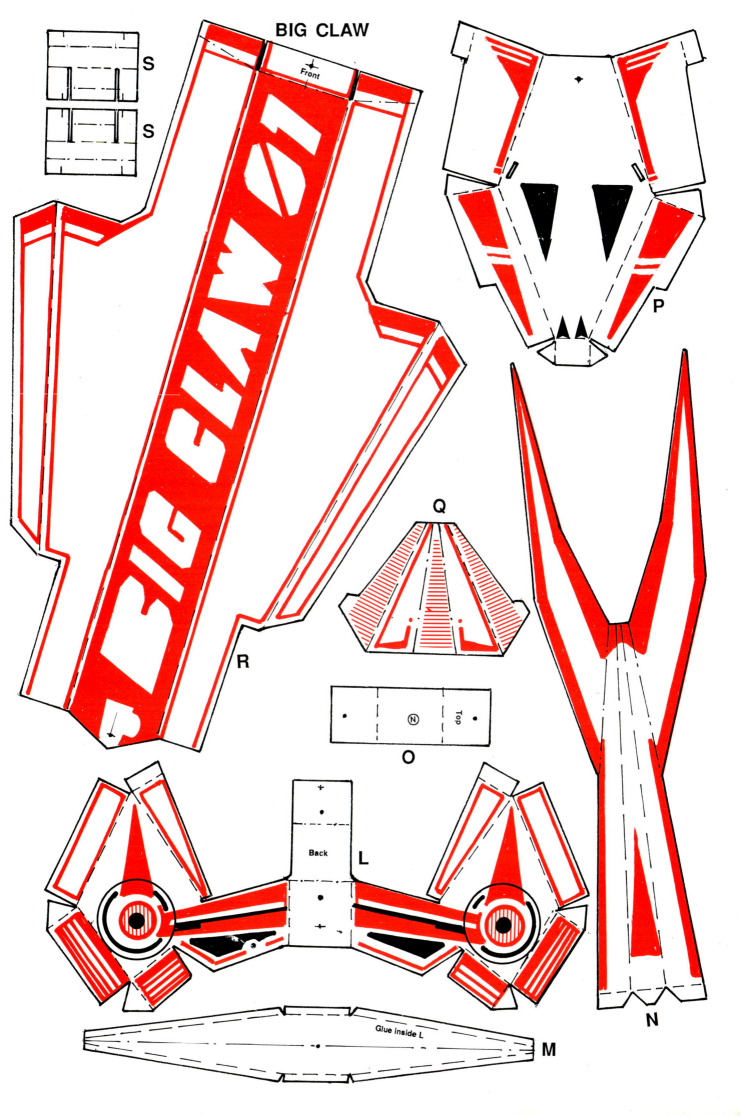